Environmental Audit

Dr. Hemant Pathak

Copyright © 2015 Dr. Hemant Pathak

All rights reserved.

ISBN: 1503363589
ISBN-13: 978-1503363588

DEDICATION

Dedicated to Shri Sainath Maharaj the all omnipotent of world the most merciful.

CONTENTS

	Foreword	5
	Glossary	8
1	Introduction	12
2	Purpose of Environmental Auditing	14
3	Phases of environmental audit	15
4	Environmental management tools	16
5	History of environmental auditing	16
6	Types of audit	17
7	Benefits of Environmental auditing	18
8	Auditing in India	20
9	Audit reporting	23
10	References	24

Foreword

Industrial concerns and local bodies should feel that they have a responsibility for abatement of pollution. Environmental Audit is an exercise of assessment to minimise the generation of wastes and pollution potential.

It has been recognised as one of the instruments for, achieving specific objectives. The procedure of an environmental statement will be introduced in local bodies, statutory authorities and public limited companies to evaluate the effect of their policies, operations and activities on the environment.

An audit statement will help in identifying and focusing attention on areas of concern, practices that need to be changed and plans to deal with adverse effects. The measures will provide better information to the public.

Environmental Audit; provides a baseline information to enable organisations to evaluate and manage environmental change, threat and risk.

This books written for academics, researchers and practitioners working in Environment field, expressed comprehensive and interdisciplinary focus on planning actions to respond to Environmental Audit.

Simply explained, Environmental Audit is an important book bringing together diverse viewpoints from Industries and state agencies and regulators, for all who wish to make a difference in how to conserve and manage our Earth's Environment.

Dr. Hemant Pathak

M.Sc. (Gold medalist), Ph. D.

Assistant Professor of Engineering Chemistry

Indira Gandhi Govt. Engineering college, Sagar, MP, India

Acronymns

- **Activity or operational audit** - an assessment of activities that may cross company departments or units

- **Audit Verification**: audits evaluate compliance to regulations or other set criteria.

- **Audit Documentation**: notes are taken during the audit and the findings recorded.

- **Audit Management**: audits can be integrated into the management system.

- **CAG:** Comptroller and Auditor General

- **Compliance audit** - the most common type of audit consisting of checks against environmental legislation and company policy;

- **Corporate audit** - an audit of the whole company and its polices, structures, procedures and practices;

- **Due diligence audit** - an assessment of potential environmental and financial risks and liabilities carried out before a company merger or site acquisition or divestiture (e.g. contaminated land remediation costs);

- **EMS**: Environmental Management System

- **External audit** - 'An assessment of the condition of the local environment, usually resulting in a State of the Environment Report (SoE or SOER)

- **health and safety audit** - an assessment of risks and contingency planning (sometimes merged with environmental auditing because of the interconnected impacts of industrial processes and hazards);

- **Internal audit** - consisting of two areas:

- **Policy Impact Assessment** - a review of the activities (objectives, services, practices and policies) of the authority' and

- **Management Audit** - a review of the procedures and structures by which environmental policies are managed by the authority'.

- **Issues audit** - an evaluation of how a company's activities relate to an environmental issue or (e.g. global pollution, energy use) or an evaluation of a specific issue (e.g. buildings, supplies);

- **O&M:** operations and maintenance

- **Objective of Audit**: information gained from the audit is reported free of opinions.

- **Periodic Audit**: audits are conducted to an established schedule.

- **Product or life cycle audit** - an analysis of environmental impacts of a product throughout all stages of its design, production, use and disposal, including its reuse and recycling

- **SAI**: Supreme Audit Institution

- **Site audit** - an audit of a particular site to examine actual or potential environmental problems;

- **Systematic Audit**: audits are carried out in a planned and methodological manner.

Glossary

Abatement	The reduction or elimination of pollution.
Affected environment	Those parts of the socio-economic and biophysical environment impacted on by the development.
Affected public	Groups, organizations, and/or individuals who believe that an action might affect them.
Alternative proposal	A possible course of action, in place of another, that would meet the same purpose and need.
Auditee	Term used to describe the person being audited. Where an organisation is being audited, it is normal to have an individual act as the auditee and coordinate responses to auditor questions, bringing in specific internal organisation experts and specialists to answer questions that the auditee cannot personally respond to.
Auditor	The auditor is the person who poses audit questions to the auditee and carries out inspections, information and document reviews.
Authorities	The national, provincial or local authorities, which have a decision-making role or interest in the proposal or activity. The term includes the lead authority as well as other authorities.
Baseline	Conditions that currently exist. Also called "existing conditions."
Baseline information	Information derived from data which: Records the existing elements and trends in the environment; and characteristics of a given project proposal
Certification Audit	A certification audit is an audit which is carried our specifically to verify that an organisation can be awarded a certificate that confirms that the organisation's environmental management systems meet the minimum requirements to formally conform to a specified ISO standard or EMAS.
Compliance Monitoring	Compliance monitoring is a continuous and systematic process to ensure that the conditions in the Environmental Management Plan (EMP) are adhered to.
Decision-maker	The person(s) entrusted with the responsibility for allocating resources or granting approval to a proposal.

Decision-making	The sequence of steps, actions or procedures that result in decisions, at any stage of a proposal.
Environment	The surroundings within which humans exist and that are made up of - i. the land, water and atmosphere of the earth; ii. micro-organisms, plant and animal life; iii. any part or combination of (i) and (ii) and the interrelationships among and between them; and iv. the physical, chemical, aesthetic and cultural properties and conditions of the foregoing that influence human health and well-being. This includes the economic, cultural, historical, and political circumstances, conditions and objects that affect the existence and development of an individual, organism or group.
Environmental Assessment (EA)	The generic term for all forms of environmental assessment for projects, plans, programmes or policies. This includes methods/tools such as EIA, strategic environmental assessment, sustainability assessment and risk assessment.
Environmental Audit	An environmental audit is a methodical examination (including tests, checks, and confirmation) of environmental procedures and practices with the view of verifying whether they comply with internal policies, accepted practices and legal requirements.
Environmental Consultant	Individuals or firms who act in an independent and unbiased manner to provide information for decision-making
Environmental Impact Assessment (EIA)	A public process, which is used to identify, predict and assess the potential environmental impacts of a proposed project on the environment. The EIA is used to inform decision-making.
Environmental Management Audit	An environmental management audit is an audit which explores the extent, nature and format of environmental management systems that are in place. It is normally carried out to evaluate operations which may be considering certification for formal EMS systems such as ISO 14000 or EMAS and require an indication of how well their existing system is functioning and what is needed to bring them up to conforming to a formal EMS system requirement.
Finding	A Finding is the term used to describe an item raised in a surveillance audit or a certification audit associated with an ISO system which requires some correction in order to ensure certification or continued certification.

Impact	The positive or negative effects on human well-being and/or on the environment.
Integrated Environmental Management (IEM)	A philosophy which prescribes a code of practice for ensuring that environmental considerations are fully integrated into all stages of the development and decision-making process. The IEM philosophy (and principles) is interpreted as applying to the planning, assessment, implementation and management of any proposal (project, plan, programme or policy) or activity - at the local, national and international level - that has a potentially significant effect on the environment.
Mitigate	The implementation of practical measures to reduce adverse impacts.
Permit Audit	A permit audit is an audit carried out, usually as a formal permit condition, to externally check the compliance of an organisation to the terms and requirements of a permit such as a water licence.
Proposal	The development of a project, plan, programme or policy. Proposals can refer to new initiatives or extensions and revisions to existing ones.
Public	Ordinary citizens who have diverse cultural, educational, political and socio-economic characteristics. The public is not a homogeneous and unified group of people with a set of agreed common interests and aims. There is no single public. There are a number of publics, some of whom may emerge at any time during the process depending on their particular concerns and the issues involved.
Scoping	The process of determining the spatial and temporal boundaries (i.e. extent) and key issues to be addressed in an environmental assessment. The main purpose of scoping is to focus the environmental assessment on a manageable number of important questions. Scoping should also ensure that only significant issues and reasonable alternatives are examined.
Screening	A decision-making process to determine whether or not a development proposal requires environmental assessment, and if so, what level of assessment is appropriate. Screening is initiated during the early stages of the development of a proposal.

Environmental audit

1. Introduction

Environmental audits are tools which can quantify an organizational environmental performance and position. An environmental audit is necessary to solve issues such as environmental legislation and pressure from customers/societies.

Environmental auditing is a methodical examination, involving analyses, tests, and confirmations of procedures and practices whose goal is to verify whether they comply with legal requirements, internal policies and accepted practices. Environmental audits will not only consist of noting differences or shortcomings that may exist at facilities, but will also acknowledge areas of exemplary performance.

It is a environmental management tool for measuring the effects of certain activities on the environment against set criteria or standards.

Environmental auditing is applicable to:

- Investigate
- Understand
- Identify

These are used to help improve existing human activities, with the aim of reducing the adverse effects of these activities on the environment. Importance of environmental audits accepted by wide range of interested parties.

Environmental audits have been conducted by SAI India for the last 25 years. This process was formalised with the introduction of specialized guidelines {MSO (Audit) 2002} for conduct of environmental audits.

An environmental auditor will study an organization's environmental effects in a systematic and documented manner and will produce an environmental audit report.

According to World Bank, environment audit is a methodical examination of environmental information about an organization, a facility or a site, to verify whether, or to what extent, they conform to specified audit criteria. The criteria may be based on local, national or global environmental standards. Thus, it is a systematic process of obtaining and evaluating information about environmental aspects.

The International Chamber of Commerce produced a definition in 1989 "A management tool comprising systematic, documented, periodic and objective evaluation of how well environmental organisation, management and equipment are performing with the aim of helping to safeguard the environment by facilitating management control of practices and assessing compliance with company policies, which would include regulatory requirements and standards applicable".

Environment auditing can encompass all types of audit, i.e., financial, compliance and performance audits.

Our environmental auditing objectives are:
- Identify and document facility compliance status. It will include evaluating patterns of deficiencies that may develop throughout the company or over time.
- Improve overall environmental performance at facilities. Regularly scheduled audits will provide an incentive for permanent resolution of environmental issues at facilities and provide a means to identify and realize continuous improvement at all operations.
- Assist facility management. In addition to verifying a facility's compliance status, audits can aid management in understanding and interpreting current or upcoming regulations or policies. They can also help identify compliance issues and cost-effective solutions as well as assist indentifying employee training needs. Further, the information provided in protocols and checklists developed by the audit group can help facilities better manage their operations.
- Increase environmental awareness.

The audit program demonstrates senior management's commitment to environmental compliance. The audit program will, by its nature, increase environmental awareness at facilities. Increasing environmental awareness will influence and involve employees at all levels of the organization. They will include proposed actions which can be taken to control, mitigate, or eliminate risks, and evaluate potential material impacts to the Company.
- Optimize Resources. Identification of environmental activities and practices, and investigation into consolidation of resources and technologies results in more efficient and cost effective management strategies.
- Provide assurance to senior management. Environmental audits provide senior management evidence that environmental affairs are being effectively managed, and that the company's exposure, including the exposure of responsible company officials, to compliance related issues

and identified hazards are minimized.

Some information of organization must for successful audits:
- Environmental policy of the audit entity.
- Financial policies relating to the environment governing the audit entity.
- Relevant rules and regulations governing the audit entity which relate to environmental compliance.
- Annual report of the audit entity.
- Administrative and financial delegation of powers of the audit entity.
- Reports of the Internal Audit of the audit entity.

The long term goal of the environmental audit program is to provide a basis for assessing and improving management systems and to identify and resolve environmental issues before they become problems, hazards, or risks.

Some of the reasons why an organisation may wish to undertake an environmental audit are:
- Pressures from environmental legislation
- Environmental liabilities and insurance costs
- Investment and decisions to buy facilities
- Detailed investigation of specific issues
- Corporate image and marketing opportunities
- Concern about the environmental impact of the organization
- Past environmental accidents

2. Purpose of Environmental Auditing

It is a systematic, objective evaluation of facility activities for a finite review period the process of determining whether our operations and practices are in compliance with regulatory requirements, Company policies and procedures, and accepted standards.

The purpose of environmental audits reflects the broadening attitudes of organisations towards environmental issues and increasing pressures from investors, insurers, consumers, and interested parties.

It is designed to:
- Verify compliance with environmental regulations, internal policies, and accepted practices.

- Valuations of environmental management systems in place, and
- Identify and measurement reasonably foreseeable risks associated with hazardous conditions attributable to our operations and prevent or mitigate such risks.

An effective corporate environmental auditing program increases environmental management effectiveness and comfort with the knowledge that the risks of potential exposure to adverse environmental issues are being addressed.

3. **Phases of environmental audit**

Each stages of environmental audit comprises a number of clearly defined objectives, with each objective to be achieved through specific actions, and these actions yielding results in the form of Outputs at the end of each phase.

An ideal audit consists of four stages:
- Planning for the audit.
- Conducting field audit.
- Audit reporting.
- Follow up review.

An environmental audit is typically undertaken in three phases:

I. Pre-audit

- Full management commitment;
- Setting overall goals, objectives, scope and priorities;
- Selecting a team to ensure objectivity and professional competence;

II. On-site audit

- On-site audit, well defined and systematic using protocols or checklists;
- Review of documents and records;
- Review of policies;
- Interviews;
- Site inspection;

III. Post-audit

- Evaluation of findings;
- Reporting with recommendations;
- Preparation of an action plan; and
- Follow-up.

4. Environmental management tools

Environmental auditing and Environmental impact assessment are environmental management tools basis on impact, effect, and significant.

EIA, attempts to predict the impact on the environment of a future action, and to provide this information to those who make the decision on whether the project should be authorized.

While Environmental auditing is carried out when a development is already in place, and is used to check on existing practices, assessing the environmental effects of current activities.

Environmental auditing covering a variety of management practices used to evaluate a company's environmental performance and gathering and evaluation of any data with environmental relevance this should actually be termed an environmental review.

5. History of environmental auditing

Environmental auditing began in the USA in the early 1970s, when a handful of industrial companies, working independently and on their own initiatives, developed environmental auditing programmes as internal management tools to help review and evaluate the status of the company operating units. It enabled managers to check compliance with

- Local environmental laws and regulations
- National environmental laws and regulations
- Corporate policies

It was also regarded as an activity useful for avoiding prosecution or civil law suits under the increasing pressures from environmental legislation.

In the world, the evolution of environmental auditing was largely due to the influence of USA subsidiary companies operating abroad. In Europe, environmental auditing began in the chemical and petrochemical industries, largely as a reflection of the intrinsic environmental hazards of these businesses, but also as a result of their involvement with American operations. Environmental auditing only became widely accepted by industry in the late 1980s as a important management tool in developed countries, and is increasingly being applied in developing countries by both foreign and local industry.

As businesses have realised the value of paying attention to environmental issues, the concept of environmental auditing itself has evolved to address wider issues than simply legal and regulatory compliance. Environmental auditing is therefore playing an increasingly common role in the management of organisations worldwide and, in some countries, governments have made the practice a legal requirement.

6. Types of audit

Environmental audit are of many types. Some audits are carried out for an entire industry, while others are for a specific site. Some audits will endeavour to investigate all aspects of environmental performance, while others are narrowly defined.

This must be a process, provide the way in which this process is utilised will depend on what the organisation wishes to achieve from the audit also leads to the use of different types of audit. An Corporate audit consist of -

    I.   Compliance viz.

- Regulatory
- EMS
- Internal standards

    II. Liability viz.

- Pre-acquisition
- Divestment
- Insurance

    III. Single Issue viz.

- Waste minimisation
- Transport etc

Other types of audits is Product audit, health and safety audit, minimisation audit, activity or operational audit.

A compliance audit aims to determine the degree of company compliance with current or prospective legislation or standards. A liability audit is usually conducted prior to buying or selling a facility/land in order to identify potential liabilities, both financial and legal.

A minimisation audit generally concentrates on a single issue, for example, waste or water, and seeks to identify ways to reduce the amount of waste produced, or water consumed. This may be the same as an issue

Audits enable the management of an organisation to see exactly what is happening within the organisation and to check the operation (or otherwise) of systems and procedures. Some environmental auditing programmes have been motivated by the occurrence of an environmental problem or incident, that is, a reactive response; others have been established in response to a desire to anticipate and head off potential problems, that is, the organisation takes a proactive stance.

The evolution of environmental auditing was largely due to the influence of USA subsidiary companies operating abroad. In Europe, environmental auditing began in the chemical and petrochemical industries, largely as a reflection of the intrinsic environmental hazards of these businesses, but also as a result of their involvement with American operations.

Environmental auditing only became widely accepted by industry in the late 1980s as a common management tool in developed countries, and is increasingly being applied in developing countries by both foreign and local industry.

7. Benefits of Environmental auditing

Environmental audits are designed to identify environmental problems, there may be widely differing reasons for undertaking them, compliance with legislation, pressure from suppliers and customers, requirements from insurers or for capital projects, or to demonstrate environmental activities to the public.

On the basis of objectives and scope of the audit. Environmental auditing benefits includes:

I. Organisations understand how to meet their legal requirements;

II. Meeting specific statutory reporting requirements;

III. Organisations can demonstrate they are environmentally responsible;

IV. Organisations can demonstrate their environmental policy is implemented;

V. Understanding environmental interactions of products, services & activities,

VI. Knowing their environmental risks are managed appropriately;

VII. Understanding how to develop and implement an EMS;

VIII. Improving environmental performance and saving money.

IX. Ensuring compliance, not only with laws, regulations and standards, but also with company policies and the requirements of an Environmental

X. Enabling environmental problems and risks to be anticipated and responses planned;

XI. To demonstrate that an organisation is aware of its impact upon the environment through providing feedback;

XII. Increased awareness amongst stakeholders; and

XIII. More efficient resource use and financial savings.

Auditing management systems provides information regarding the choice of Environmental Auditor. Environmental Auditors should have personal attributes, such as ethics, open-mindedness, perceptiveness and tact.

They should understand audit principles, procedures and techniques, as well as having gained experience through conducting audits. They should know the subject matter they are auditing against and how this applies to different organisations.

Audit Team Leaders should be able to plan and resource effectively, have good communication and leadership skills. Preferably Environmental Auditors should complete training and have attained an appropriate level of education. A good Auditor should have adequate skills and experience.

When seeking an external Auditor consideration could be given to the skills outlined above. Exemplar Global Environmental Auditors have completed training and have met a minimum certification standard. Depending on the auditing requirements consideration could be given to determining whether the Auditor needs to be certified by additional organisations.

8. Auditing in India

The Supreme Audit Institution in India is headed by the Controller and Auditor General (CAG) of India that is a constitutional authority. The Supreme Audit Institution of India through its various field offices has been conducting compliance audit of government laws, rules and legislations as well as performance audit of government programmes and schemes. A Regional Training Institute in Mumbai has been designated as the nodal training center to impart training to officers and staff of SAI India on environmental auditing.

The CAG of India derives his mandate from Articles 148 to 151 of the Indian Constitution.

The CAG's (Duties, Powers and Conditions of Service) Act, 1971 prescribes functions, duties and powers of the CAG. While fulfilling his constitutional obligations, the CAG examines various aspects of government expenditure and revenues.

The audit conducted by CAG is broadly classified into Financial, Compliance and Performance Audit.

Environmental audit by SAI India is conducted within the broad framework of Compliance and Performance Audit.

I. Environment protection in India

The Ministry of Environment & Forests (MOEF) authorized for the planning, promotion, co-ordination and overseeing the implementation of environmental and forestry programmes.

The Ministry is also the Nodal agency in the country for the United Nations Environment Programme (UNEP) also main agency for implementation of environment programmes.

The principal activities undertaken by Ministry of Environment & Forests consist of :
• Conservation & survey of flora, fauna, forests and wildlife;

- Prevention & control of pollution;

- Afforestation and regeneration of degraded areas; and

- Protection of environment, in the frame work of legislations.

   Major policy initiatives by Ministry of Environment and Forests include:
- National Environment Policy, 2006;

- National Conservation Strategy and Policy Statement on Environment and Development, 1992;

- Policy Statement for Abatement of Pollution;

- National Forest Policy etc,.

Ministry of Environment and Forests has enacted various rules/ regulations/notifications for control of water pollution, air pollution, environment protection, animal welfare, wildlife etc.

This laid down broad guidelines to enable SAI India's auditors to examine whether the auditee institutions gave due regard to the efforts of promulgating sustainability development and environmental concerns, where warranted.

With a view to bring in a focused attention and consolidate the approach for better outputs and infusing new techniques and tools, SAI India designated the office of the Principal Director of Audit (Scientific Departments) as the nodal office for Environmental Audit. This office undertakes exclusively, environmental audits of central government programmes, all over the country.

As a result of having a specialized office and a training institute, SAI India has a vast pool of audit professionals equipped with techniques of environmental auditing.

SAI India has conducted more than 100 specialised environmental audits over the last 25 years.

The current work in the area of environmental auditing being done by SAI India includes:

- Preparing "Green Office Guidelines" which is a write-up for guiding SAI India offices all over the country to reduce the detrimental effects of office operation on the environment by more sustainable and efficient use of office resources. This write-up had also been adopted by the Ministry of Environment of Forests and is being circulated to all central government offices in India.

- Assisting the Ministry of Environment and Forests in preparing a policy for the effective management of Waste in India.

As representative of SAI India, Principal Director (Scientific Departments) is a member of a committee which will evolve policy and strategy for the better management of waste in India based on the recommendations made by SAI India in its Audit Report on "Management of Waste in India".

- Preparing guidelines for "4th E—Integrating Environmental concerns in Auditing" which is guidance on incorporating environmental concerns in all kinds of audit.

4. Environmental audits carried out by C&AG of India

More than 100 environmental audits (compliance and performance) have been carried out by SAI India over the last 20 years.

The audits can be divided into 5 categories—

(i) Air issues
(ii) Water issues
(iii) Waste
(iv) Biodiversity
(v) Environment Management System.

Air issues In 2002, audit of Air Pollution/ Vehicular emissions/Industry emissions was conducted in 23 States of India which showed that poor implementation and monitoring of the Air Pollution Control Act led to increase in pollution levels.

Some other audits on this issue are:

- Ineffective pollution control in Thermal Power Stations of Bihar State Electricity Board, 2005.
- Pollution Control by Transport Department of Mizoram, 2006.

Environmental auditing is a systematic, documented, periodic and objective process in assessing an organization's activities and services in relation to:

Assessing compliance with relevant statutory and internal requirements

- Facilitating management control of environmental practices
- Promoting good environmental management
- Maintaining credibility with the public

- Raising staff awareness and enforcing commitment to departmental environmental policy
- Exploring improvement opportunities

Establishing the performance baseline for developing an Environmental Management System (EMS)

9. Audit reporting

A audit reports should consist of :

(a) A contents list.

(b) An executive summary indicating the principal findings.

(c) An introduction highlighting the scope and purpose of the audit and report.

(d) An assessment of performance against previously agreed criteria, highlighting strengths, weaknesses and identifying non-compliances.

(e) Reference to corrective items from previous audits.

(f) List of action items and recommendations

10 . References

BBC News online (Wednesday 18 July 2007) Cement Firm's Pollution Fine Cut.

Crump A (1991) Dictionary of Environment and Development: People, Places, Ideas and Organisations. Earthscan, London.

Dagg S (2005) C108 Environmental Auditing. Module prepared for the Distance Learning Programme, Imperial College London.

Gensburg LJ, Pantea C, Fitzgerald E, Stark A, Hwang S-A, Kim N (2009) Mortality among former Love Canal residents. Environmental Health Perspectives **117**(2) 209–216.

Humphrey N, Hadley M (2000) Environmental Auditing. Palladian Law Publishing Ltd, Bembridge, Isle of Wight.

Hunt D, Johnson C (1995) Environmental Management Systems. McGraw Hill, London.

International Chamber of Commerce (1989) Environmental Auditing. June 1989, ICC Publication No 468, International Chamber of Commerce (ICC), Paris.

International Chamber of Commerce (1991) ICC Guide to Effective Environmental Auditing. ICC Publication No 483, International Chamber of Commerce (ICC), Paris.

Meadows DH, Meadows DL, Randers J, Behrens III WW (1972) The Limits to Growth. Universe Books, New York.

Smets H (1988) The cost of accidental pollution. Industry and Environment **11**(4) 28–33.

Thetimes100 (2010) Business Objectives, Planning and Stakeholders. http://www.thetimes100.co.uk/theory/theory--business-objectives-planning-stakeholders--321.php

UN (2002) World Summit on Sustainable Development (WSSD). Johannesburg 26 August to 4 September 2002, United Nations (UN). http://www.un.org/events/wssd/

ABOUT THE AUTHOR

Dr. Hemant Pathak held positions as Assistant Professor in the department of chemistry, Govt. Indira Gandhi Engineering College, Sagar, MP, India. He had extensive experience in teaching, research and administrative management.

Dr. Pathak received his Ph.D. degree in chemistry from Dr. Hari Singh Gour Central University, Sagar, India and M.Sc. Gold medalist from Jiwaji University, Gwalior. He has published 24 books and more than 70 research papers in reputed International and National journals and received several awards. He is a member of editorial boards and reviewer boards of several international journals and societies. His area of specialization includes Engineering Chemistry, Energy audits and Environmental Pollution management.

www.ingramcontent.com/pod-product-compliance
Lightning Source LLC
Chambersburg PA
CBHW081823170526
45167CB00008B/3516